I0446196

GROWING YOUR OWN FOOD

Dr. Kimberly Carlos

TABLE OF CONTENT

SELF SUFFICIENCY MADE EASY FOR BEGINNERS AND SENIORS LOOKING TO REAP WHAT THEY SOW

CHAPTER ONE

The Journey Begins

Introduction

In an era dominated by bustling cities, fast food, and global supply chains, there's a growing movement towards rediscovering our connection to the earth and the food we consume. In an age where "farm-to-table" and "locally sourced" have become buzzwords, there's an innate desire to return to our roots, to reclaim the simple yet profound act of growing our own food. In this increasingly complex and fast-paced world, there's something undeniably grounding, therapeutic, and empowering about cultivating your vegetables, herbs, and fruits in your own backyard, balcony, or community garden. This is a journey that offers not just a harvest of fresh, nutritious produce but a multitude of physical, mental, and emotional rewards.

"The Journey Begins," the opening chapter of our book, "Growing Your Own Food," is the gateway to this captivating voyage of self-sufficiency and sustainability. It introduces us to the profound and multifaceted world of home gardening, beckoning us to embark on a journey that will transform the way we view food, nature, and ourselves.

The Benefits of Growing Your Own Food

So why embark on this journey in the first place? The reasons are as diverse as the people who choose to grow their own food. For some, it's a response to concerns about the quality and safety of commercially grown produce. The awareness of pesticides, herbicides, and long-distance transportation of food raises questions about what exactly we're putting on our tables. For others, it's about the joy of watching a tiny seedling transform into a flourishing plant, or the satisfaction of knowing that the tomatoes on your sandwich came from your own garden.

But the benefits extend far beyond just the produce itself. This journey reconnects us with the earth, fostering a deep appreciation for the seasons, the soil, and the ecosystem that supports our sustenance. It's an opportunity to nurture patience, resilience, and an understanding of the delicate balance of nature. Gardening is a therapeutic endeavor, known to reduce stress, improve mental well-being, and provide a sense of accomplishment that's hard to replicate in our modern, high-speed world.

Understanding the Environmental Impact

Growing your own food isn't only about personal well-being; it's also a powerful step toward reducing your ecological footprint. As we'll explore in this chapter, large-scale industrial agriculture is a significant contributor to climate change, habitat destruction, and loss of biodiversity. By taking control of your food supply and adopting sustainable gardening practices, you become a vital part of the solution. Your garden becomes a mini-haven for pollinators, a carbon sink, and a model of responsible land stewardship.

In "Guide to Growing Your Own Food," we delve into the numerous advantages of growing your own food, from enhancing your health and well-being to making a positive impact on the environment. This chapter sets the stage for a voyage that can be as personal or as communal as you choose, whether you're tending a single potted plant on your apartment balcony or contributing to a neighborhood garden initiative.

It's time to embark on a journey that promises fulfillment, sustainability, and a deeper connection to the world around us. " Guide to Growing Your Own Food " will provide you with the knowledge, inspiration, and motivation to get your hands in the soil, to nurture the miracle of life, and to savor the fruits of your labor. The adventure of growing your own food begins here, and the possibilities are as boundless as the garden that awaits you.

Introduction to Growing Your Own Food

In a world dominated by the convenience of supermarkets and the allure of pre-packaged meals, the act of growing one's food may seem like a distant relic from the past. Yet, the age-old practice of nurturing plants, tending to the soil, and harvesting your own sustenance is experiencing a revival in recent years. The movement to grow your own food has gained momentum as people from all walks of life are discovering the many rewards and benefits of this deeply fulfilling endeavor.

Growing your own food isn't just a means of putting fresh, organic produce on your table, although that's undoubtedly one of its most enticing aspects. It's a journey of self-sufficiency, a connection to the land, and an opportunity to regain control over what you eat. It's a statement of independence, a commitment to sustainability, and a return to the rhythms of nature. This introduction serves as a gateway to the world of home gardening, a world that promises not only nourishment but also personal growth, environmental stewardship, and a profound sense of fulfillment.

Why You Should Grow Your Food

The advantages of growing your own food are diverse and deeply rewarding. First and foremost, it offers an unparalleled source of fresh, nutrient-rich produce. There's a special kind of satisfaction that comes from plucking a ripe tomato, a crisp cucumber, or a bundle of fragrant basil from your own garden and transforming it into a meal. You can savor the flavors of your labor and know precisely where your food comes from.

Beyond the tangible rewards, gardening is a therapeutic pursuit. It reconnects you with the natural world, allowing you to escape the demands and distractions of modern life. The act of nurturing plants, of feeling the soil beneath your fingers, has been proven to reduce stress, improve mental well-being, and instill a sense of accomplishment that few other hobbies can match.

Understanding the Environmental Impact

The benefits of growing your own food extend beyond personal well-being. They reach deep into the realm of environmental responsibility. Large-scale, industrial agriculture has long been associated with environmental issues such as deforestation, water pollution, and the emission of greenhouse gases. When you embark on the journey of growing your own food, you take a step towards sustainability. Your garden can become a haven for pollinators, a small but essential refuge for wildlife, and a testament to the principles of conservation and responsible land use.

"The Journey Begins" is an exploration of the possibilities that unfold when you embrace the art of growing your own food. Whether you have a spacious backyard, a small urban balcony, or even just a windowsill, there's a place for you in this movement. It's a journey filled with lessons in patience, resilience, and the delicate balance of the natural world. It's a voyage that can be as personal or as communal as you choose, from tending a single potted plant to joining a community garden initiative.

In the pages that follow, we will delve into the practical aspects of gardening, from preparing the soil to nurturing your plants and harvesting the fruits of your labor. "Growing Your Own Food" is not just a how-to guide; its an invitation to embrace a lifestyle that's in harmony with the earth, your health, and your spirit. Its an invitation to embark on a journey of self-discovery and sustainability, and the adventure begins with the decision to grow your own food.

The Benefits of Growing Your Own Food

In a world increasingly characterized by fast food, processed meals, and the constant rush of modern life, the act of growing your own food is a transformative journey that offers an array of profound benefits. Beyond the tangible rewards of fresh, homegrown produce, the practice of cultivating your own fruits, vegetables, and herbs provides a range of advantages that touch upon physical health, mental well-being, sustainability, and a deeper connection to the environment.

1. Nutritional Superiority: Perhaps the most evident benefit of growing your own food is the ability to harvest produce at its peak of ripeness and nutritional value. Supermarket fruits and vegetables are often picked prematurely and transported over long distances, resulting in a loss of nutrients. In contrast, homegrown crops can be harvested when fully mature, ensuring that you consume the freshest and most nutritious food possible.

2. Control Over What You Eat: In an era when concerns about pesticides, genetically modified organisms (GMOs), and synthetic additives in food are on the rise, growing your own food provides a powerful sense of control. You decide what goes into your garden, from the seeds or plants you select to the cultivation methods you employ. This control over your food supply contributes to peace of mind and better health.

3. Environmental Impact: Industrial agriculture is known for its detrimental environmental consequences, including habitat destruction, deforestation, and greenhouse gas emissions. When you grow your own food, you actively reduce your ecological footprint. Your garden can become a haven for pollinators, support biodiversity, and, through sustainable gardening practices, mitigate environmental damage. This act of small-scale farming fosters an appreciation for responsible land stewardship.

4. Therapeutic Benefits: Gardening is not merely a practical endeavor; it is also a therapeutic pursuit. The act of tending to plants, feeling the earth in your hands, and observing the life cycles of your crops provides a powerful mental and emotional respite from the fast pace and stresses of modern life. Gardening is known to reduce stress, improve mood, and enhance overall mental well-being.

5. Self-Sufficiency and Independence: Growing your own food instills a sense of self-sufficiency and independence. It's a statement of empowerment, a realization that you can nurture and provide for your basic needs. This sense of autonomy is especially appealing in a world where so much of our daily existence relies on external systems and resources.

6. Connection to Nature: As you immerse yourself in the practice of growing food, you forge a profound connection to the natural world. You become attuned to the changing seasons, the intricacies of soil health, and the delicate interplay of life in your garden. This connection enriches your understanding of the environment and fosters a deeper respect for the planet.

"The Benefits of Growing Your Own Food" are multifaceted, offering not only the pleasures of fresh and delicious produce but also the profound rewards of improved health, a smaller environmental footprint, and a more meaningful connection to the earth. As you delve into the world of gardening, you'll discover that it's not just about what you grow in your garden; it's about what you grow within yourself and the positive impact you make on the world around you.

Understanding the Environmental Impact

In an age where sustainability and ecological responsibility are increasingly at the forefront of public consciousness, understanding the environmental impact of growing your own food is paramount. Home gardening isn't just about cultivating your favorite vegetables; it's a practice that can significantly contribute to a healthier planet by mitigating the adverse effects of industrial agriculture. Let's delve into how your efforts in the garden can make a positive difference for the environment.

1. Reduced Carbon Footprint: One of the most glaring issues with conventional agriculture is the extensive transportation network required to move food from the farm to your plate. The food on your grocery store shelves may have traveled hundreds or even thousands of miles. Growing your own food drastically reduces this carbon footprint. Your produce doesn't need to be transported, thereby reducing greenhouse gas emissions and lessening your contribution to climate change.

2. Preservation of Biodiversity: Large-scale monoculture farming often leads to the reduction of biodiversity. By growing a variety of crops in your garden, you can help protect and promote the biodiversity of local flora and fauna. Moreover, your garden can serve as a sanctuary for beneficial insects, pollinators, and other wildlife, contributing to a healthier and more balanced ecosystem.

3. Reduced Chemical Usage: Conventional farming frequently involves the application of synthetic pesticides, herbicides, and fertilizers that can leach into the environment, leading to soil and water pollution. In contrast, many home gardeners opt for organic or low-impact cultivation methods, reducing the harmful effects of these chemicals. Furthermore, practicing crop rotation and companion planting can naturally help control pests and diseases in your garden.

4. Water Conservation: Industrial agriculture is notorious for its heavy water consumption. However, as a home gardener, you have greater control over your water use. Techniques such as drip irrigation, mulching, and rainwater harvesting can help conserve water and reduce the strain on local water supplies. By optimizing your water management, you contribute to sustainable resource use.

5. Soil Health: Industrial farming practices can degrade soil quality through excessive tilling and the overuse of chemicals. In contrast, many home gardeners focus on enriching and maintaining their soil through composting and other organic practices. This promotes healthier, more fertile soil that, in turn, sequesters carbon and reduces erosion.

6. Promotion of Sustainable Practices: As you embark on your journey of growing your own food, you become an advocate for sustainable and responsible land use. Your garden serves as a model for friends and neighbors, inspiring them to explore their own gardening ventures and adopt environmentally friendly practices. Collectively, this encourages a broader shift towards more sustainable food production.

Understanding the environmental impact of growing your own food underscores the profound potential for positive change. As a home gardener, you're not merely cultivating vegetables; you're contributing to a more sustainable, ecologically responsible future. Your garden is a testament to the power of small-scale, mindful agriculture and a reflection of your commitment to nurturing both your well-being and the health of the planet.

Assessing Your Motivation and Goals

Before you dig your first garden bed or plant your inaugural seedling, it's essential to take a step back and consider your motivation and goals for growing your own food. Understanding why you want to embark on this journey and what you hope to achieve will guide your decisions and efforts throughout the gardening process. Your motivations can vary widely, but identifying them can help you tailor your gardening approach to align with your desires and priorities.

1. Health and Nutrition: Many individuals are drawn to growing their own food because of a desire for healthier, more nutritious options. If your motivation is primarily health-related, you likely aim to cultivate pesticide-free, organic produce that is rich in essential vitamins and minerals. Gardening allows you to have complete control over the quality and safety of your food, ensuring that it is free from harmful chemicals and preservatives.

2. Environmental Consciousness: For those motivated by environmental concerns, growing your own food is an eco-conscious choice. Your goal may be to reduce your carbon footprint by minimizing the transportation of food, support local biodiversity, conserve water, and reduce the use of harmful chemicals. Understanding these goals will lead you to adopt sustainable practices in your garden and make choices that benefit the planet.

3. Self-Sufficiency and Resilience: Some people are drawn to gardening by the desire to become more self-sufficient and resilient in the face of economic and environmental uncertainties.

If this is your motivation, your goal may be to decrease reliance on external food systems and become more self-reliant. Your garden becomes a source of food security, providing a measure of independence in an increasingly interconnected world.

4. Therapeutic and Stress Relief: Gardening is renowned for its therapeutic benefits. If your motivation is centered on stress relief and personal well-being, your goal may be to create a peaceful, meditative space. The act of nurturing plants, connecting with nature, and witnessing the growth of your garden can have a profound impact on your mental health and overall quality of life.

5. Financial Savings: Growing your own food can also be a means of reducing grocery expenses. If you are motivated by saving money, your goal may be to harvest bountiful crops that reduce your food costs. Careful planning and efficient gardening practices can help you achieve these financial objectives.

6. Education and Skill Development: Gardening offers endless opportunities for learning and skill development. If you are motivated by the pursuit of knowledge and new abilities, your goal may be to develop expertise in gardening, experiment with different techniques, and deepen your understanding of plant biology, soil science, and sustainable agriculture.

As you assess your motivation and goals for growing your own food, it's important to recognize that these motivations are not mutually exclusive. They can overlap and evolve over time. What's crucial is that your gardening journey is a reflection of your values and aspirations. By understanding your reasons for embarking on this journey, you'll be better equipped to make informed choices, set realistic goals, and tailor your gardening practices to meet your unique objectives.

Whether you're motivated by health, environmental concerns, self-sufficiency, or the joy of learning, your garden becomes a canvas for personal growth and positive change. By aligning your goals with your motivations, you'll not only yield a fruitful harvest but also cultivate a profound sense of satisfaction and fulfillment.

Planning Your Garden Space

One of the most crucial steps in the journey of growing your own food is the thoughtful and strategic planning of your garden space. Your garden's layout, size, and organization will significantly influence the success and productivity of your homegrown produce. In this phase, careful consideration and forethought are essential, as they can save you time, effort, and resources in the long run.

1. Location: Begin your garden planning by selecting the optimal location. The amount of sunlight, wind exposure, and proximity to water sources are all critical factors to consider. Most vegetables and fruits require at least 6 to 8 hours of direct sunlight each day. Additionally, ensure that your chosen site is easily accessible for regular maintenance and harvesting.

2. Garden Layout: The layout of your garden is vital to its efficiency and aesthetic appeal. Popular garden layouts include traditional rows, raised beds, container gardens, and even vertical gardening. Consider your space and gardening goals when selecting a layout that best suits your needs. Raised beds, for instance, are excellent for smaller spaces and can be easier to maintain.

3. Crop Rotation: To promote healthy soil and prevent pest and disease buildup, it's crucial to plan for crop rotation. This means not planting the same family of vegetables in the same spot in successive years. Keep records of what you plant each year, and design your garden layout to accommodate this rotation.

4. Companion Planting: Understanding which plants thrive when grown together and which ones should be kept apart is a key aspect of garden planning. Companion planting can help deter pests, improve pollination, and enhance overall plant health. For example, planting basil near tomatoes can improve the latter's flavor and health.

5. Soil Preparation: Proper soil preparation is the foundation of a successful garden. Perform a soil test to assess the nutrient content and pH level of your soil. Depending on the results, you may need to amend the soil with organic matter, such as compost, to improve its fertility and structure.

6. Plant Selection: Choose crops that are well-suited to your region, taking into account factors like your local climate, the length of your growing season, and the soil type. Selecting varieties that are naturally resistant to common pests and diseases can reduce the need for chemical interventions.

7. Space Management: Take into consideration the space requirements of different crops. Some vegetables, like sprawling pumpkins or vining cucumbers, may need more room to grow, while compact plants like lettuce can be interplanted in smaller spaces. Proper spacing between plants is essential for good air circulation and healthy growth.

8. Accessories and Infrastructure: Consider any accessories or infrastructure you'll need. This might include trellises for climbing plants, a watering system, a composting area, and paths for easy navigation. Such additions can make your garden more functional and aesthetically pleasing.

9. Long-Term Planning: Gardening is a dynamic process, and your garden space should accommodate your long-term goals. Plan for potential expansion, changes in crop selection, and improvements in infrastructure as you gain experience and refine your gardening skills.

Effective planning of your garden space can significantly increase your chances of a successful and productive harvest. It ensures that you make the most of your available space, resources, and time. By carefully considering the location, layout, and other important factors, you set the stage for a bountiful garden that not only provides delicious, homegrown food but also becomes a source of pride and joy.

Choosing the Right Crops for Your Region

When embarking on the journey of growing your own food, one of the most critical decisions you'll make is selecting the right crops for your specific region. The choice of crops should be based on your local climate, soil conditions, and growing season. Adapting your garden to the natural conditions of your area is a fundamental step toward a successful harvest and a rewarding gardening experience.

1. Know Your Hardiness Zone: The first step in selecting the right crops is to determine your hardiness zone.

Hardiness zones, as defined by the United States Department of Agriculture (USDA) and similar organizations in other countries, help you understand the average minimum winter temperatures in your region. This information is essential for choosing plants that can thrive in your climate.

2. Consider Your Climate: Beyond hardiness zones, consider the broader climate of your region. Factors such as temperature ranges, humidity, rainfall, and the length of your growing season all play a crucial role in plant selection. Some crops, like tomatoes and peppers, require a warm climate and a long growing season, while others, such as spinach and kale, are better suited to cooler conditions.

3. Soil Analysis: Your local soil conditions also significantly impact plant growth. Conduct a soil test to determine the pH level and nutrient content of your soil. This analysis will help you select crops that are well-suited to your soil type and provide guidance on any necessary soil amendments.

4. Water Availability: Understanding your region's water availability is essential for choosing the right crops. If you live in an area with regular rainfall, you may have more flexibility in your crop selection.

In contrast, arid regions may require drought-tolerant varieties, and areas with poor drainage may benefit from raised beds or container gardening.

5. Growing Season: The length of your growing season is a critical factor in selecting crops. Some regions have shorter growing seasons due to frost dates, while others can enjoy year-round gardening. It's crucial to choose crops that can mature within your specific growing window to ensure a successful harvest.

6. Local Expertise: Seek advice from local gardeners, agricultural extension services, or gardening clubs in your area. Local expertise can provide valuable insights into the specific challenges and opportunities of your region. Fellow gardeners can recommend crops that have thrived in your locale and offer tips for success.

7. Crop Diversity: To maximize your garden's productivity and minimize the risk of crop failure, diversify your crop selection. Include a mix of vegetables, fruits, and herbs that are well-suited to your region. Crop diversity can also enhance soil health and help manage pests and diseases.

8. Experiment and Learn: Gardening is a dynamic and evolving practice. Don't be afraid to experiment and learn from your experiences. Over time, you'll develop a deeper understanding of what works best in your region, enabling you to fine-tune your crop selection each season.

By choosing the right crops for your region, you set the stage for a thriving garden and a successful journey into self-sufficiency and sustainability. This thoughtful approach ensures that your efforts are aligned with the unique conditions of your area, and it maximizes the chances of a bountiful harvest of homegrown, fresh, and delicious produce that's well-suited to your local environment.

Creating a Budget for Your Garden

Embarking on the journey of growing your own food is a rewarding endeavor, but it's essential to approach it with financial prudence. Crafting a budget for your garden is a fundamental step in planning for a successful and sustainable gardening experience. Whether you have a small balcony garden or a spacious backyard plot, a well-thought-out budget will help you manage expenses, allocate resources wisely, and enjoy the fruits of your labor without financial stress.

1. Initial Expenses: Start your garden budget by considering the initial expenses. These may include the cost of seeds or seedlings, soil amendments, garden tools, and essential infrastructure such as raised beds or containers. Research prices and make a list of the items you'll need to get your garden started.

2. Recurring Costs: Gardens require ongoing care, and recurring expenses can add up. Consider costs such as water, fertilizers, pest control measures, and mulch. Also, plan for garden maintenance tools and equipment like hoses, pruners, and gloves.

3. Plant Selection: Your choice of crops can significantly impact your budget. Some plants, like certain herbs and leafy greens, are relatively low-cost to grow, while others, such as exotic or space-intensive varieties, can be more expensive. Determine which plants align with your budget and preferences.

4. Soil and Compost: Healthy soil is the foundation of a productive garden. Invest in soil tests to assess the quality of your soil and identify necessary amendments.

While quality soil can be an upfront cost, it's a long-term investment that can lead to higher yields and less reliance on external fertilizers.

5. Garden Infrastructure: If your garden requires additional infrastructure, such as trellises, irrigation systems, or garden structures, include these expenses in your budget. These investments can improve efficiency and reduce long-term costs.

6. Educational Resources: Consider allocating part of your budget for educational resources. Gardening books, courses, and workshops can enhance your knowledge and skills, ultimately saving you time and money by helping you avoid common mistakes.

7. Seeds and Seedlings: While starting plants from seeds is often more budget-friendly, some gardeners prefer to purchase seedlings to save time and effort. Assess your priorities and budget for either option accordingly.

8. Pest and Disease Management: Pest and disease management is a part of every garden. Budget for the necessary tools and supplies, such as natural pesticides or beneficial insects, to help protect your plants.

9. Harvesting and Preservation: Don't forget to budget for tools and supplies related to harvesting and food preservation. This may include canning jars, drying racks, and other preservation equipment if you plan to store your harvest.

10. Contingency Fund: Gardening can be unpredictable, and unexpected expenses may arise. To avoid financial strain, allocate a portion of your budget as a contingency fund for unforeseen costs or emergencies.

11. Record Keeping: Maintaining records of your expenses is crucial for tracking your budget and making informed decisions in subsequent seasons. Invest in a budgeting tool or create a simple spreadsheet to monitor your spending.

12. Long-Term Planning: Think beyond the current season and consider long-term investments, such as perennial crops or establishing fruit trees, which can provide a continuous source of food and reduce future expenses.

Creating a budget for your garden is a practical and empowering exercise that ensures your gardening journey is both fulfilling and financially sustainable. By carefully considering all potential costs, allocating resources wisely, and practicing financial discipline, you can enjoy the pleasures of homegrown produce while maintaining a healthy balance in your budget. Gardening not only enriches your table but also provides a sense of financial prudence and a deeper connection to your food and its sources.

CHAPTER TWO

Preparing the Ground

In the realm of growing your own food, one of the most fundamental and transformative phases is preparing the ground. It's a process that can be likened to setting the stage for a grand production.

The soil, the very foundation of your garden, becomes the canvas upon which the story of your plants will unfold. To prepare the ground is to lay the groundwork for a thriving and abundant harvest, and it involves far more than just digging a few holes and sowing some seeds.

"The ground" in gardening encompasses the physical earth that supports your plants, but it also extends to the metaphorical ground of your commitment, knowledge, and aspirations.

This introduction explores the multifaceted journey of preparing the ground for your garden, delving into the physical, the practical, and the profound aspects of this essential endeavor.

The Soil's Hidden Riches

One of the most critical aspects of preparing the ground is understanding the soil beneath your feet. Soil is not merely dirt; it's a complex ecosystem teeming with life, a repository of nutrients, and a dynamic environment that influences the success of your garden. Different soils possess unique qualities, from the sandy soils that drain quickly to the loamy soils that hold moisture and nutrients.

Understanding your soil type, its pH level, and nutrient content is the first step in successful gardening. Soil testing can reveal these secrets, helping you determine whether your soil requires amendments to achieve the ideal conditions for plant growth. Whether through the addition of organic matter, compost, or specific nutrients, preparing the ground involves transforming your soil into a nurturing environment for your crops.

The Ritual of Preparation

Preparing the ground is a ritual, a connection to the land that has been repeated by generations of gardeners. It's a symphony of physical labor, of turning over the earth, breaking up clods, and removing rocks.

It's the gentle caress of seeds finding their new home and the cover of a blanket of soil to protect and nourish them. This ritual connects you to the timeless cycle of life, growth, and sustenance.

In this introductory chapter, we'll explore the practical aspects of soil preparation, including techniques for clearing the ground, amending the soil, and creating garden beds. You'll learn about the tools you'll need, from shovels to rakes, and the importance of working efficiently and diligently. We'll discuss the concept of raised beds and container gardening, which can be particularly valuable in areas with poor or contaminated soil.

The Profound Connection

Beyond the practicality of soil preparation, there's a deeper connection to be forged with the land. Gardening is not merely a physical act but a spiritual and emotional one as well.

The act of nurturing the soil, the promise of new life emerging from your efforts, and the anticipation of a bountiful harvest all contribute to a profound sense of connection and fulfillment.

"The ground" in preparing the ground is a metaphor for the foundation of your journey. It's about connecting with the earth, embracing the seasons, and recognizing that you are a steward of the land.

This introduction sets the stage for the chapters to come, where we'll explore the physical and emotional aspects of preparing the ground for your garden. It's a journey that involves hard work, patience, and a deep respect for the earth and its cycles. As you embark on this path, you'll not only cultivate your garden but also cultivate a connection to the natural world and to the timeless wisdom of those who have tilled the soil before you.

Soil Health and Soil Testing

In the world of gardening and agriculture, soil is the unsung hero, the foundation upon which all plant life depends. The health of your soil is a critical determinant of the success of your garden, as it directly affects the growth, nutrition, and vitality of your plants.

Understanding soil health and the practice of soil testing are fundamental components of responsible and productive gardening.

The Significance of Soil Health

Soil health refers to the overall condition of the soil, including its physical, chemical, and biological properties. Healthy soil is not merely a lifeless medium for holding plants; it's a dynamic ecosystem teeming with beneficial microorganisms, nutrients, and organic matter. Here's why soil health is of paramount importance:

1. Nutrient Supply: Healthy soil provides essential nutrients like nitrogen, phosphorus, and potassium, which are necessary for plant growth. These nutrients are made available to plants through the intricate web of soil microorganisms.

2. Water Retention and Drainage: Soil health influences the soil's ability to hold moisture and drain excess water. Balanced soil structure prevents waterlogging and ensures adequate hydration for your plants.

3. Pest and Disease Resistance: A robust, healthy soil ecosystem can naturally suppress harmful pests and diseases, reducing the need for chemical interventions.

4. Root Growth: Strong root development is supported by healthy soil, leading to more extensive root systems that can access nutrients and water effectively.

5. Carbon Sequestration: Healthy soil can store significant amounts of carbon, contributing to carbon sequestration and mitigating climate change.

The Practice of Soil Testing

Soil testing is the process of analyzing a soil sample to determine its nutrient content, pH level, and other essential characteristics. This diagnostic tool is invaluable for gardeners, farmers, and land managers because it provides precise information about your soil's current condition. Here's what you can learn from a soil test:

1. pH Level: Soil pH measures its acidity or alkalinity. Different plants have specific pH requirements, so knowing your soil's pH allows you to select plants that will thrive in your garden.

2. Nutrient Levels: Soil tests reveal the levels of essential nutrients such as nitrogen, phosphorus, and potassium. You can then adjust your fertilization practices to meet your plants' needs.

3. Organic Matter: The amount of organic matter in your soil affects its structure, water-holding capacity, and nutrient retention. Soil testing can indicate whether you need to add organic matter to improve soil health.

4. Cation Exchange Capacity (CEC): CEC is a measure of your soil's ability to hold onto and exchange nutrients with plant roots. Soils with higher CEC can better retain and deliver nutrients to plants.

5. Micronutrients: Soil tests can identify deficiencies or excesses of essential micronutrients like iron, zinc, and manganese, enabling you to address any imbalances.

6. Toxic Elements: In some cases, soil tests can identify the presence of harmful elements like heavy metals or contaminants that may affect plant and human health.

Applying Soil Test Results

Once you've conducted a soil test and received your results, it's crucial to apply this knowledge to your gardening practices. You can adjust your choice of plants, amend your soil with the right nutrients or organic matter, and optimize your fertilization regimen based on the test's recommendations.

Incorporating the concept of soil health and regular soil testing into your gardening routine is a transformative step toward becoming a responsible and successful gardener. It promotes sustainability, enhances plant growth, reduces resource waste, and contributes to the overall well-being of your garden ecosystem. Soil is not just the ground beneath your plants; it's a living, breathing entity that plays a vital role in nurturing and sustaining the natural world. Understanding and caring for your soil is a profound and rewarding aspect of the journey of growing your own food.

Composting and Organic Matter

In the world of sustainable gardening and agriculture, composting is often heralded as a powerful alchemical process, turning what might be considered waste into a precious resource. At its heart, composting is a celebration of organic matter's cyclical journey, transforming kitchen scraps, yard waste, and other biodegradable materials into nutrient-rich, humus-like soil conditioner. The value of composting extends far beyond your garden, as it embodies the principles of recycling, environmental stewardship, and the cultivation of life in the soil.

The Magic of Composting

Composting is a remarkable process driven by the decomposition of organic matter, orchestrated by an orchestra of microorganisms, fungi, and beneficial insects. This complex and symbiotic relationship transforms raw materials like food scraps, leaves, grass clippings, and coffee grounds into a nutrient-dense, dark, and crumbly substance known as compost.

Here's what makes composting magical:

1. Waste Reduction: Composting diverts organic waste from landfills, reducing the emission of methane, a potent greenhouse gas. Instead, it transforms this waste into a valuable resource.

2. Soil Enrichment: Compost is a powerhouse of nutrients, improving soil structure, water-holding capacity, and aeration. It enhances the fertility and health of your garden soil, providing essential organic matter and nutrients to your plants.

3. Natural Pest and Disease Suppression: Compost supports the growth of beneficial microorganisms that can outcompete harmful pathogens, reducing the risk of plant diseases.

4. Reduced Erosion: Amended with compost, soil is better equipped to resist erosion by improving its texture and structure.

5. Water Conservation: Compost enhances soil's ability to hold moisture, reducing the need for frequent watering and helping plants survive droughts.

The Art of Composting

Creating compost requires a bit of art and science. To embark on this journey, you'll need a compost bin or pile, a balance of green and brown materials, and the right environmental conditions:

1. Green Materials: These are rich in nitrogen and include items like fruit and vegetable scraps, grass clippings, and coffee grounds. Green materials provide protein and energy for the decomposers.

2. Brown Materials: Brown materials, rich in carbon, include dry leaves, straw, wood chips, and newspaper. They provide fiber and structure for the compost.

3. Aeration: Composting microorganisms need oxygen to thrive. Turn your compost pile or use aeration tools to provide ample oxygen.

4. Moisture: The compost pile should be as damp as a wrung-out sponge. Too much water can lead to anaerobic conditions, while too little can slow down decomposition.

5. Layering: Alternate layers of green and brown materials in your compost pile. This balance ensures proper decomposition and minimizes odors.

6. Size Matters: Chop or shred larger materials to speed up decomposition, as smaller pieces have more surface area for microorganisms to work on.

Pitfalls to Avoid

While composting is a relatively simple process, a few common pitfalls are worth noting:

1. Inadequate Aeration: Without proper oxygen levels, decomposition can slow down or become anaerobic, leading to unpleasant odors.

2. Inappropriate Materials: Avoid adding diseased plants, meat, dairy, or pet waste to your compost, as they can introduce pathogens or pests.

3. Improper Balance: A balance between green and brown materials is essential for efficient decomposition.

4. Neglecting Moisture: A compost pile that's too dry or too wet can lead to decomposition problems.

Composting and the incorporation of organic matter into your garden are acts of stewardship, embracing the principles of sustainability, recycling, and environmental responsibility. By harnessing the magic of decomposition, you create a self-renewing cycle that enriches your garden soil, supports plant health, and minimizes waste. Your garden thrives, the environment benefits, and you become an active participant in the circle of life that sustains us all. Composting is both a practical and philosophical journey, enriching not only your soil but also your connection to the earth and its abundant offerings.

Garden Bed Preparation

In the world of gardening, the phrase "bed preparation" might sound as though you're getting ready for a nap, but it's quite the opposite. Garden bed preparation is a critical and active phase in the journey of growing your own food. This is where you lay the foundation for your plants' growth, ensuring they have the right environment to flourish. Whether you're working with traditional ground beds, raised beds, or containers, the principles of bed preparation are the same and will guide you toward a successful harvest.

The Soil's Promise

At the heart of garden bed preparation is the soil itself. Your soil is not merely a passive medium; it's a dynamic, living ecosystem. To prepare your garden bed is to shape this environment, ensuring it's rich in nutrients, well-structured, and teeming with life. Here are the essential steps to follow:

1. Clear the Area: Begin by clearing the area of weeds, rocks, debris, and any existing vegetation that may compete with your crops for nutrients and space. It's the equivalent of preparing a clean canvas for your gardening masterpiece.

2. Soil Testing: Conduct a soil test to understand your soil's pH, nutrient levels, and any deficiencies. The results will guide your decisions about soil amendments.

3. Amend the Soil: Based on your soil test, add organic matter and nutrients to your soil. Compost, well-rotted manure, and other organic materials are invaluable for improving soil structure and fertility.

4. Till or Turn the Soil: To blend the amendments into your existing soil, you can till the soil using a mechanical tiller or, for smaller gardens, turn the soil by hand using a spade or garden fork. This process helps distribute nutrients and organic matter evenly.

5. Create Raised Beds: If you're opting for raised beds, construct or fill the bed frames with a combination of quality garden soil and compost. Raised beds can provide better drainage, reduce compaction, and enable earlier planting in cooler climates.

6. Mulch: Apply a layer of mulch to the surface of your garden bed. Mulch helps retain moisture, suppress weeds, and regulate soil temperature.

Plan for Spacing

While preparing your garden bed, it's essential to consider plant spacing and layout. This planning ensures that your crops have adequate room to grow, access to sunlight, and good air circulation, which can reduce the risk of diseases. Factors to consider include the mature size of your plants, the recommendations on seed packets or plant labels, and any specific spacing requirements for your chosen crops.

Protection and Maintenance

Once your garden bed is prepared and your plants are in the ground, the work isn't over. It's important to maintain your garden beds throughout the growing season. Regular weeding, watering, and monitoring for pests and diseases are essential to ensure your plants remain healthy and productive. You may also need to apply additional organic matter or nutrients as the growing season progresses to keep your soil fertile.

Garden bed preparation is an investment in the success of your garden. A well-prepared garden bed not only provides an optimal environment for plant growth but also minimizes the challenges and frustrations that can arise from poor soil quality and improper spacing. It's a practice that brings you closer to the heart of gardening, where the earth and your efforts come together to produce a bountiful harvest. As you prepare your garden beds, you're shaping the stage for your gardening journey, setting the scene for a thriving, flourishing spectacle of life, color, and flavor that will unfold in the months to come.

Raised Beds and Container Gardening

In the world of home gardening, not every space is a sprawling expanse of fertile soil, and not every gardener has the luxury of a traditional in-ground garden. That's where raised beds and container gardening come into play, offering versatile solutions for growing your own food in limited spaces or challenging conditions. These methods open up a world of possibilities, making it possible to cultivate a vibrant, productive garden no matter the size of your outdoor area or the quality of your existing soil.

Raised Beds: Gardening on a Pedestal

Raised beds are like the elegant crowning jewels of the garden. They are contained garden plots that sit above the ground, often enclosed by wooden frames or other materials.

There are several advantages to this method:

1. Better Soil Control: One of the key benefits of raised beds is the ability to create custom soil mixes. You can tailor the soil to suit the specific needs of your crops, ensuring optimal drainage and nutrient levels.

2. Improved Drainage: Raised beds often have better drainage than in-ground gardens, making them suitable for regions with heavy rainfall or clayey soil that doesn't drain well.

3. Warmer Soil: Raised beds warm up more quickly in the spring, allowing for earlier planting and extended growing seasons in cooler climates.

4. Reduced Soil Compaction: Since you don't walk on the soil in raised beds, there's less soil compaction, which can impede root growth and water infiltration.

5. Accessibility: The height of raised beds is advantageous for gardeners with physical limitations. You can comfortably garden without bending or kneeling.

Container Gardening: Space-Saving Versatility

Container gardening is a versatile approach that involves growing plants in pots, containers, or other confined spaces. It's particularly valuable when you have limited outdoor space or need to adapt to challenging environments. Here's what makes container gardening appealing:

1. Adaptability: You can place containers on patios, balconies, rooftops, or even indoors by a sunny window, making it ideal for urban or small-space gardening.

2. Soil Control: Just like with raised beds, container gardeners have complete control over the soil mix. You can tailor it to suit the specific needs of your plants.

3. Mobility: Containers are portable. You can move them to optimize sunlight or protect plants from harsh weather conditions.

4. Pest and Disease Control: Container gardening can help isolate plants from garden pests and diseases, reducing the risk of infestations.

5. Aesthetics: Containers add a touch of aesthetics to your outdoor space. They come in various sizes, shapes, and materials, allowing you to create visually appealing arrangements.

Tips for Success

Whether you choose raised beds or container gardening, success lies in your approach:

1. Quality Soil: Invest in high-quality soil or create a custom soil mix that meets your plants' needs. Good soil is the foundation of your garden's success.

2. Proper Drainage: Ensure that containers and raised beds have proper drainage to prevent waterlogged soil, which can harm plant roots.

3. Watering Care: Container gardens often require more frequent watering, as containers can dry out faster. Be diligent in maintaining appropriate moisture levels.

4. Choose the Right Plants: Select plants that are suitable for your space, considering factors like sunlight, space, and the size of your containers or raised beds.

5. Regular Maintenance: Just like in-ground gardens, container gardens and raised beds need regular attention, including weeding, feeding, and pest management.

Raised beds and container gardening are not only solutions for small spaces but also opportunities for creativity and innovation. They demonstrate that, even in the most challenging circumstances, a thriving garden is attainable. Whether you have a small balcony or a patch of rocky soil, these methods empower you to harness the joy of homegrown produce and bring the bounty of the garden to your doorstep.

Soil Amendments and Fertilization

In the world of gardening, soil is the canvas upon which the story of plant life unfolds, and soil amendments and fertilization are the tools that help paint a vibrant and bountiful picture. These practices are fundamental to nurturing the health and fertility of your garden's soil, ensuring that your plants receive the nutrients they need to thrive. By enhancing the soil's structure, nutrient content, and biological activity, you can create an environment where plants flourish, diseases are minimized, and yields are maximized.

Soil Amendments: Building Blocks of Fertility

Soil amendments are organic or inorganic materials added to soil to improve its physical and chemical properties. These amendments enhance soil structure, water-holding capacity, drainage, and nutrient retention. Some common organic amendments include compost, well-rotted manure, and leaf mold, while inorganic amendments may include vermiculite or perlite. Here's why soil amendments are indispensable:

1. Improved Soil Structure: Amendments break up compacted soil and enhance its structure, allowing for better root penetration, air circulation, and water infiltration.

2. Nutrient Enhancement: Organic amendments gradually release essential nutrients as they decompose, enriching the soil with valuable elements that are crucial for plant growth.

3. Water Retention: Organic matter, like compost, acts as a sponge, retaining moisture in the soil. This is particularly valuable during dry periods, as it reduces the need for frequent watering.

4. Aeration: Soil amendments create pockets of air in the soil, providing oxygen to plant roots and encouraging beneficial soil microorganisms.

5. Biological Activity: Organic matter fosters the growth of beneficial microorganisms, earthworms, and other organisms that aid in nutrient cycling and disease suppression.

Fertilization: Nourishing Your Plants

While soil amendments enrich the soil, fertilization directly provides essential nutrients to your plants. Fertilizers contain concentrated sources of nitrogen (N), phosphorus (P), and potassium (K), along with other micronutrients. Proper fertilization practices are critical for promoting healthy growth, flowering, and fruiting in your plants.

Here's what you should know about fertilization:

1. Understanding Nutrient Needs: Different plants have varying nutrient requirements. By knowing the specific needs of your crops, you can select the appropriate fertilizer to address deficiencies and optimize growth.

2. Balanced Fertilization: A balanced fertilizer provides all the essential nutrients in the right proportions. The label on the fertilizer package will typically indicate the N-P-K ratio (e.g., 10-10-10), representing the percentage of each nutrient.

3. Application Timing: Fertilizer should be applied at the right time in the plant's growth cycle. For instance, nitrogen is typically needed in greater quantities during the vegetative growth stage, while phosphorus supports flowering and fruiting.

4. Applying Sufficiently: Applying fertilizer according to the recommended rates is crucial. Under-fertilizing can lead to nutrient deficiencies, while over-fertilizing may harm the plants and the environment.

5. Organic vs. Synthetic Fertilizers: Both organic and synthetic fertilizers have their advantages. Organic fertilizers release nutrients slowly and improve soil structure, while synthetic fertilizers offer precise nutrient delivery. Gardeners often choose a combination of both for optimal results.

The Balanced Approach

A balanced approach to soil amendments and fertilization involves starting with healthy soil through the addition of organic matter, tending to the soil's specific needs based on regular soil tests, and providing targeted fertilization for your plants. This approach promotes a thriving garden that's not only lush with greenery and delicious produce but also sustainable and environmentally responsible. It's a testament to the gardener's art of nurturing the earth, cultivating life, and embracing the wonder of the growing cycle that plays out year after year in the soil beneath your feet.

Pest and Weed Management Strategies

In the pursuit of growing your own food, one of the key challenges that gardeners face is managing pests and weeds. These unwelcome visitors can quickly become a source of frustration and hinder the success of your garden.

However, effective pest and weed management strategies can help you maintain a thriving garden without resorting to harmful chemicals. Here are some eco-friendly and sustainable approaches to manage these common garden adversaries.

Pest Management Strategies

1. Companion Planting: Companion planting involves strategically placing plants in proximity to one another to enhance growth and repel pests. For example, planting marigolds alongside your vegetables can deter aphids, while basil can help repel tomato hornworms.

2. Beneficial Insects: Encourage the presence of beneficial insects like ladybugs, lacewings, and parasitic wasps, which are natural predators of common garden pests. You can attract them by planting flowers that provide nectar and pollen.

3. Crop Rotation: Rotate your crops annually to disrupt the life cycles of pests and reduce the likelihood of infestations. Different crops have different pest vulnerabilities, so this strategy can help break the cycle.

4. Handpicking: Regularly inspect your plants for pests and remove them by hand. This method works particularly well for larger insects like caterpillars and beetles.

5. Neem Oil and Insecticidal Soaps: These natural substances can be used as insect repellents. Neem oil has pesticidal properties and can be effective in controlling a range of garden pests.

6. Barrier Methods: Physical barriers, such as row covers or netting, can protect your plants from insects while allowing air, light, and moisture to pass through.

Weed Management Strategies

1. Mulch: Applying organic mulch, like straw, wood chips, or compost, to the soil surface can help suppress weed growth by blocking sunlight and preventing weeds from taking root.

2. Weeding by Hand: Regularly inspect your garden and remove weeds by hand when they are small and easier to pull. Be sure to remove the entire root to prevent regrowth.

3. Hoeing: A hoe can be used to slice through weed seedlings just below the soil surface, disrupting their growth. Be cautious not to disturb your crops while hoeing.

4. No-Till Gardening: No-till gardening involves minimal soil disturbance, which can help reduce weed germination and preserve soil structure. It's an excellent approach for weed management in larger garden plots.

5. Cover Crops: Planting cover crops like clover or rye during the off-season can prevent weed growth by outcompeting them for space and nutrients.

6. Weed Barriers: Landscape fabric or cardboard can be used as weed barriers around plants or in pathways. Ensure that soil or mulch covers these materials to create a more aesthetic appearance.

7. Organic Herbicides: Organic herbicides containing acetic acid or citrus oil can be used to target weeds without harming other plants. Use them with caution, as they may affect beneficial insects.

Remember that a holistic approach is often the most effective. Encouraging biodiversity in your garden, practicing good sanitation, and maintaining healthy soil are all part of an integrated pest and weed management strategy.

Be patient and observant in your garden, and learn to identify common garden pests and weeds. Prevention is often the most effective strategy, so regularly monitor your garden to address issues before they become overwhelming.

Irrigation and Watering Techniques

Water is the lifeblood of your garden, and efficient irrigation and watering techniques are essential for nurturing healthy, thriving plants.

Gardening success is closely tied to providing the right amount of water at the right time. The proper methods can conserve water, prevent overwatering, and ensure that your plants receive the hydration they need.

Let's explore some effective irrigation and watering techniques to help you maintain a lush and productive garden.

Drip Irrigation: Precision and Efficiency

Drip irrigation is a highly efficient method that delivers water directly to the root zones of your plants. It uses a network of tubes, pipes, and emitters to provide a slow and controlled release of water.

This approach offers several advantages:

1. Water Conservation: Drip irrigation reduces water wastage by delivering water precisely where it's needed, minimizing runoff and evaporation.

2. Reduced Weeds: Since the water is delivered to the plants and not the surrounding soil, weed growth is less encouraged.

3. Less Disease Risk: Watering leaves and foliage can create conditions for diseases to thrive. Drip irrigation avoids wetting the plant's aerial parts.

4. Automated Control: Drip systems can be automated with timers, ensuring that your plants receive water even when you're not around.

Soaker Hoses: Gentle and Uniform

Soaker hoses are porous hoses that release water slowly along their length. They are an excellent choice for garden beds, providing gentle and uniform watering.

Key advantages include:

1. Even Coverage: Soaker hoses distribute water evenly, preventing overwatering in some areas and underwatering in others.

2. Minimal Water Loss: Like drip irrigation, soaker hoses minimize water loss to evaporation and runoff.

3. Ease of Use: Soaker hoses are easy to install and can be moved or adjusted as needed.

4. Convenient for Rows: They are particularly effective for rows of plants or garden beds with a linear layout.

Overhead Sprinklers: Broad Coverage

Overhead sprinklers deliver water in a widespread pattern, making them suitable for large areas or lawns. While they're less water-efficient than drip or soaker systems, they have their merits:

1. Convenience: Sprinklers are easy to set up and cover large areas, making them suitable for lawns and landscapes.

2. Cooling Effect: On hot days, sprinklers can provide a cooling effect for your plants, reducing stress.

3. Watering Seeds: Sprinklers are ideal for germinating seeds, as they provide even moisture distribution for newly planted areas.

Hand Watering: Personal Touch

Hand watering, using a hose or watering can, provides the personal touch that connects you to your plants. It's especially effective for potted plants or specific areas that need extra attention. Key benefits include:

1. Visual Inspection: Hand watering allows you to closely observe your plants, providing an opportunity to check for pests, diseases, or other issues.

2. Precision: You can control the amount and location of water, ensuring individual plants receive the care they need.

3. Flexibility: Hand watering is flexible and can be tailored to the needs of different plants or areas.

Watering Tips for Success

1. Water Early or Late: Water in the morning or late afternoon to reduce evaporation and give plants ample time to dry before evening.

2. Avoid Foliage: Aim for soil-level watering to avoid wetting leaves and reducing the risk of diseases.

3. Consistency: Establish a regular watering schedule to maintain consistent moisture levels for your plants.

4. Deep Watering: Ensure that water penetrates the root zone to encourage deep and strong root growth.

5. Mulch: Apply mulch to your garden beds to help retain soil moisture, reduce weeds, and regulate soil temperature.

6. Adjust for Seasons: Be mindful of changing weather conditions and adjust your watering frequency accordingly.

Successful irrigation and watering are fundamental to a flourishing garden. By choosing the right techniques for your plants and garden layout, you can conserve water, promote plant health, and enjoy the beauty and abundance of a well-hydrated garden.

Whether you opt for drip irrigation, soaker hoses, or good old-fashioned hand watering, mastering these techniques is the key to cultivating a garden that thrives throughout the growing season.

CHAPTER THREE

Planting and Cultivating

In the grand symphony of gardening, planting and cultivating are the crescendos, the moments when the seeds of your labor and dreams take root in the earth. These are the acts where your nurturing spirit meets the boundless potential of the natural world. It's a journey that involves more than just putting plants in the ground; it's about fostering life, embracing the cycles of growth, and experiencing the awe-inspiring transformations that occur as your garden unfolds.

The Art of Planting: Nurturing Beginnings

Planting is a ritual of hope and promise. It's the moment when you take those seeds or young seedlings, cradle them gently in your hands, and entrust them to the soil. It's a practice that connects you with the profound cycles of life, the promise of growth, and the tangible act of transformation. When you plant a seed, you're not just burying it in the ground; you're planting a piece of your vision for a garden teeming with life and nourishment.

Planting is a study in diversity. In your garden, each plant has its unique needs and idiosyncrasies. It's your role as a gardener to understand the specific requirements of each species and ensure they find a place where they can thrive. Whether you're planting vibrant flowers, hearty vegetables, or lush trees, you're orchestrating a harmonious blend of colors, shapes, and scents that will fill your garden with vibrancy.

The Cultivation Continuum: Care, Patience, and Growth

Cultivating is the act of tending to your garden, of nurturing your plants through their journey of growth. This practice involves much more than simply watering and weeding. It's an art that requires you to pay close attention to your garden's unique needs, to understand the language of the soil and the rhythm of the seasons.

Cultivation is a demonstration of patience. Gardening is a practice of delayed gratification, where the rewards come gradually, where the little seedlings you planted eventually grow into strong, fruitful plants. It's a reminder that the most precious things in life often take time to develop, to mature, and to reveal their true potential.

But it's not just about waiting; it's about active care. You'll find yourself adjusting your garden's environment, providing the right nutrients, protecting it from pests and diseases, and pruning and training your plants to grow in desired shapes and patterns.

Cultivation is the act of being attentive to the garden's changing needs, responding to the challenges of changing weather and adapting to the evolving requirements of your plants.

The Garden as a Canvas: Your Creative Sanctuary

Planting and cultivating are about more than just growing food or creating a beautiful landscape. They're an expression of creativity and a sanctuary for the soul.

Your garden is a canvas where you can explore the interplay of colors, shapes, and textures. It's a retreat where you can find solace and connect with the natural world.

In your garden, you're both an artist and a steward of the earth, shaping the environment to reflect your vision while respecting the cycles of life that sustain us all.

We will delve into the art of planting and cultivating, exploring the techniques, knowledge, and practices that transform your aspirations into reality. We will cover the specifics of seed planting, transplanting, and providing the optimal conditions for growth.

We will discuss the care and nurturing of your garden as it matures, embracing the challenges and the rewards of the gardening journey.

Through this, you will not only cultivate your plants but also nurture a profound connection to the natural world, the seasons, and the timeless wisdom that has guided gardeners for generations.

Starting from Seeds or Seedlings

The decision to begin your gardening journey by starting from seeds or seedlings is a pivotal one, as it sets the stage for the entire process of cultivating your garden.

Each approach has its advantages and considerations, and the choice ultimately depends on your goals, resources, and gardening preferences.

Starting from Seeds: The Miracle of Life

Embarking on your gardening adventure from seeds is like witnessing the miracle of life unfold before your eyes. It's a practice deeply rooted in the ancient traditions of agriculture and horticulture, and it offers several compelling benefits:

1. Variety: Seeds provide access to a vast array of plant varieties and cultivars, allowing you to select the specific traits, flavors, and characteristics that align with your garden goals.

2. Cost-Efficiency: Seeds are generally more affordable than buying established seedlings or plants. You can grow many plants from a single packet of seeds, making it a budget-friendly option.

3. Sustainability: Starting from seeds is an eco-friendly choice, as it reduces plastic waste associated with purchasing nursery-grown plants and minimizes the carbon footprint of transporting plants.

4. Full Lifecycle Experience: Growing plants from seeds offers a deeper understanding of a plant's lifecycle, from germination to maturity. It's a profoundly educational and rewarding experience.

However, starting from seeds also requires patience, time, and careful attention. You must create the right conditions for germination, which typically include providing adequate light, moisture, and temperature control. It's important to note that some plants may have a longer lead time when grown from seeds, as they need to reach maturity before they yield harvestable crops.

Starting from Seedlings: A Head Start

Opting for seedlings, or young plants, provides a head start in your gardening endeavor. These are typically young plants that have already germinated and developed into small, ready-to-plant seedlings. This approach offers several advantages:

1. Time-Saving: Seedlings allow you to skip the initial germination phase, giving you a head start on the growing season. You can plant them when the weather is suitable, potentially yielding an earlier harvest.

2. Easier On Beginners: Starting with established seedlings can be less intimidating for novice gardeners who may not yet have the skills or experience to successfully germinate seeds.

3. Predictability: Seedlings give you a clear idea of what your mature plant will look like, making it easier to plan your garden layout.

4. Quick Harvest: If you're eager to enjoy a harvest sooner rather than later, seedlings can provide a more immediate gratification.

However, purchasing seedlings can limit your variety choices compared to starting from seeds.

Additionally, it can be more expensive, as each plant must go through the initial germination phase, which involves resources and time on the part of the nursery or grower.

Finding the Balance

The choice between starting from seeds or seedlings ultimately depends on your individual preferences, goals, and circumstances. Many gardeners find that a combination of both approaches works best.

For some plants, starting from seeds may be more practical, while for others, especially those with longer growing seasons or those requiring specific environmental conditions, beginning with seedlings may be the way to go.

In either case, whether you're nurturing seeds into life or tending to young plants, the essence of gardening lies in the journey of caring for and connecting with your garden.

It's a deeply rewarding and transformative experience, regardless of whether you choose seeds or seedlings to launch your adventure.

Proper Plant Spacing and Layout

In the intricate choreography of gardening, proper plant spacing and layout are the dance steps that ensure harmony, balance, and the well-being of your garden.

The arrangement of your plants in your garden bed is more than just an aesthetic choice; it's a key factor in optimizing growth, health, and productivity.

Getting it right involves understanding the unique requirements of your crops, maximizing the use of space, and considering long-term growth and maintenance.

The Importance of Proper Plant Spacing

1. Air Circulation: Spacing your plants adequately allows for better air circulation, reducing the risk of fungal diseases. Plants with enough space can also dry out more efficiently after rain or watering.

2. Light Exposure: Proper spacing ensures that each plant receives an adequate amount of sunlight. This is crucial for photosynthesis and healthy growth.

3. Nutrient Availability: Plants that are too close together may compete for nutrients. Adequate spacing ensures that each plant has access to the nutrients it needs to thrive.

4. Reduced Pest Pressure: Crowded plants can provide hiding spots and pathways for pests. Proper spacing can help deter the buildup of pests.

5. Weed Suppression: Well-spaced plants allow for the use of mulch and ground cover to suppress weeds, reducing competition for resources.

Determining Plant Spacing:

The specific plant spacing you should use varies depending

on the plant species and its growth habits. Consult seed packets or plant labels for spacing recommendations. Here are some general guidelines:

1. Small Plants: Most small or dwarf varieties need 6 to 12 inches of spacing between each plant. This includes many herbs and low-growing annuals.

2. Medium Plants: Crops like lettuce, spinach, and some flowering perennials typically require 12 to 18 inches between plants.

3. Large Plants: Medium-sized vegetables like tomatoes, peppers, and bush beans need 18 to 24 inches of spacing. Taller, sprawling plants may require even more space.

4. Vining Plants: Vining crops, such as cucumbers, melons, and squash, can take up significant space and require 24 to 36 inches or more between each plant. Consider trellising or vertical gardening for these crops to maximize space.

5. Trees and Shrubs: Trees and shrubs vary greatly in size, so spacing depends on the specific variety. Always consult planting recommendations or guidelines from a nursery.

Layout Considerations

In addition to plant spacing, the overall layout of your garden beds or containers is essential for efficient use of space and aesthetic appeal. Some key considerations include:

1. Rows vs. Blocks: You can arrange plants in traditional rows or use a block planting method. Rows are suitable for larger gardens, while blocks can maximize space and reduce water waste.

2. Companion Planting: Consider companion planting principles, which involve placing plants together that benefit each other in terms of pest control, pollination, or nutrient sharing.

3. Crop Rotation: Plan your layout with crop rotation in mind to prevent soil depletion and disease buildup.

4. Paths and Accessibility: Ensure that your layout includes pathways for easy access, maintenance, and harvesting.

5. Aesthetics: The layout should be visually appealing, with colors, textures, and plant heights considered.

Remember that your garden is an evolving ecosystem, and

as the growing season progresses, plants will change in size and shape. Some may require staking or support, while others might need additional space. Regular maintenance and adaptation to your layout will be necessary to accommodate the dynamic nature of your garden.

Proper plant spacing and layout are at the heart of a productive and visually pleasing garden. By taking the time to plan and arrange your plants thoughtfully, you're not only optimizing their growth but also creating a beautiful and well-organized garden that you can enjoy throughout the growing season.

Crop Rotation for Healthy Soil

In the intricate dance of gardening, crop rotation is a well-choreographed move that not only optimizes plant growth but also promotes soil health.

It's a practice deeply rooted in the wisdom of generations of gardeners, and it serves as a natural and sustainable method to maintain fertile, disease-resistant, and well-balanced soil.

By rotating your crops season after season, you can unlock a world of benefits for your garden.

Preventing Soil Depletion

Crop rotation is essential for preventing soil depletion, a condition where the soil loses essential nutrients and fertility over time due to the repeated cultivation of the same crops. Different plants have different nutrient requirements, and when the same crops are grown in the same place year after year, they continuously draw upon the same nutrients. This leads to imbalances in the soil's nutrient content, with some elements being depleted and others accumulating to excess.

By rotating crops, you introduce variety into the soil's diet. For instance, nitrogen-fixing legumes, like beans and peas, can follow heavy nitrogen feeders such as corn. This practice allows the soil to recover and replenish the nutrients that were heavily consumed during the previous growing season. As a result, you'll have healthier soil that can support robust plant growth and higher yields.

Disease and Pest Control

Crop rotation can also play a pivotal role in disease and pest control. Many plant-specific pests and pathogens have a limited host range. When you rotate crops, you disrupt the life cycles of these pests and diseases, reducing their impact on your garden.

For example, if your garden has been plagued by tomato blight in one season, rotating your crops the following year can break the cycle.

You can plant a different family of plants, like beans or cucumbers, in that area. This reduces the risk of the disease spreading, as it won't find its preferred host plants in the same location.

Enhancing Soil Structure

Crop rotation can also help improve soil structure. Different crops have varying root structures and depths, which means they loosen the soil to varying degrees. Deep-rooted crops like carrots can break up compacted soil layers, improving aeration and water infiltration.

In contrast, shallow-rooted crops, such as lettuce, don't penetrate the soil as deeply but contribute to surface organic matter.

By rotating between deep and shallow-rooted plants, you promote soil health by maintaining a good balance of aeration and organic matter throughout the soil profile.

Long-Term Sustainability

Crop rotation is a practice that fosters long-term sustainability in your garden. It ensures that the soil remains fertile, well-structured, and capable of supporting healthy plant growth. This means you won't have to rely as heavily on synthetic fertilizers or pesticides, as your garden will naturally resist disease and maintain balanced nutrient levels.

Planning Your Crop Rotation

An effective crop rotation plan depends on the size of your garden, the number of crops you want to grow, and the specific requirements of each plant. You can divide your garden into sections and rotate crops through those sections on a seasonal basis. Be sure to keep a record of what you planted in each section to track the rotation. The key is to avoid planting the same crop or its close relatives in the same location for at least two to three years.

Crop rotation is a fundamental practice for maintaining healthy soil and optimizing your garden's productivity. By diversifying your plantings, you contribute to the resilience and fertility of your garden, nurturing a living ecosystem that thrives season after season. It's a sustainable, time-honored practice that pays rich dividends in the form of vibrant, productive, and disease-resistant gardens.

Mulching for Weed Suppression

In the age-old battle between gardeners and weeds, mulching stands as a steadfast ally. It's a practice that not only enhances the aesthetic appeal of your garden but also plays a critical role in weed suppression. Mulch, whether organic or inorganic, creates a protective barrier that shades the soil, regulates moisture, and discourages weeds from taking root. This simple yet effective technique is a cornerstone of sustainable gardening, providing a range of benefits beyond weed control.

Weed Suppression through Mulch

1. Shading the Soil: One of the primary ways mulch suppresses weeds is by shading the soil. A thick layer of mulch effectively blocks sunlight from reaching the soil surface. This prevents weed seeds from germinating and stops established weeds from photosynthesizing, effectively smothering them.

2. Reducing Soil Temperature: In addition to shading, mulch helps regulate soil temperature. Weed seeds often require specific temperature ranges to germinate, and mulch can keep the soil cooler, discouraging the germination of many weed seeds.

3. Moisture Regulation: Mulch helps retain soil moisture by reducing water evaporation. Weed seeds often need consistently moist soil to germinate. By preserving moisture, mulch can thwart their germination efforts.

4. Physical Barrier: Mulch creates a physical barrier between the soil and the environment, making it more challenging for weed seeds to come into contact with the soil and establish roots.

Types of Mulch

There are various types of mulch, each with its unique advantages for weed suppression:

1. Organic Mulch: Organic mulch includes materials like straw, wood chips, leaves, grass clippings, and compost. Organic mulch not only suppresses weeds but also decomposes over time, enriching the soil with nutrients.

2. Inorganic Mulch: Inorganic mulch, such as plastic or landscape fabric, is long-lasting and offers excellent weed suppression. However, it does not contribute to soil improvement as organic mulch does.

3. Living Mulch: Ground cover plants like clover or creeping thyme can serve as living mulch. These plants not only suppress weeds but also add beauty to your garden.

4. Mulch Combinations: Combining different types of mulch can provide both weed suppression and improved soil quality. For instance, you can place landscape fabric or cardboard under wood chips to maximize weed control while allowing the organic mulch to break down and enrich the soil.

Applying Mulch for Weed Suppression:

To effectively suppress weeds with mulch, follow these steps:

1. Prepare the Soil: Remove existing weeds and cultivate the soil before applying mulch. A weed-free start sets the stage for successful weed suppression.

2. Apply a Thick Layer: For optimal weed control, apply a layer of mulch 2-4 inches deep. Ensure the layer is thick enough to block sunlight from reaching the soil.

3. Maintain Mulch Depth: Mulch can break down over time, so it's essential to replenish it periodically to maintain its weed-suppressing effectiveness.

4. Leave Space around Plants: Keep mulch several inches away from plant stems to prevent moisture-related diseases and rot. Create a ring of mulch-free soil around each plant.

Year-Round Benefits

Mulching for weed suppression is not a one-time task but a year-round practice. It helps reduce the time and effort required for weeding, allowing you to focus on nurturing your garden and enjoying its beauty and bounty.

Moreover, the organic mulch decomposes over time, improving soil quality and supporting healthy plant growth.

This is a sustainable and ecologically sound method for achieving a harmonious garden that thrives with minimal weed interference.

Pruning and Training Your Plants

Pruning and training your plants are essential techniques in the gardener's toolkit, allowing you to shape, direct, and maintain healthy and productive plants. These practices involve selectively removing or encouraging specific parts of a plant to achieve desired outcomes, such as enhanced growth, better fruit production, and improved aesthetics. Properly pruning and training your plants not only helps them reach their full potential but also contributes to garden health and beauty.

Pruning: Shaping and Promoting Growth

Pruning is the art of selectively cutting or removing parts of a plant, such as branches, stems, or leaves, to achieve specific goals. Pruning is essential for various reasons:

1. Size and Shape Control: Pruning allows you to maintain a plant's size and shape, preventing it from becoming too unruly or invasive in your garden.

2. Disease Prevention: Removing dead or diseased branches can prevent the spread of diseases throughout the plant and to nearby ones.

3. Air Circulation and Sunlight: Thinning out dense growth through pruning can improve air circulation and increase sunlight penetration, reducing the risk of fungal diseases and enhancing photosynthesis.

4. Fruit Production: Pruning can stimulate increased fruit production by redirecting a plant's energy into producing fruit instead of excessive foliage.

5. Rejuvenation: Pruning can rejuvenate old or overgrown plants by removing old, unproductive branches and encouraging new growth.

When pruning, it's essential to use sharp, clean tools and make precise cuts.

The timing of pruning varies by plant type, so it's crucial to research the specific requirements of the plants in your garden to ensure you're not pruning at the wrong time of year.

Training: Guiding Growth for a Purpose

Training plants involves manipulating their growth patterns to achieve specific objectives, such as encouraging vines to climb, guiding the growth of fruit trees, or creating ornamental shapes. Some common training techniques include:

1. Espalier: This method involves training fruit trees or shrubs to grow flat against a wall or trellis, creating a two-dimensional, decorative form that also maximizes fruit production.

2. Topiary: Topiary is the art of sculpting plants into ornamental shapes through pruning and training. This technique is often used for hedges or decorative garden features.

3. Trellising and Staking: Training climbing plants like peas, beans, and tomatoes to grow on trellises or stakes can maximize space and light exposure.

4. Bonsai: Bonsai involves pruning and training trees or shrubs to grow in small containers while maintaining their ornamental appearance.

5. Training Vines: Vining plants like grapes, roses, and wisteria can be trained to grow on arbors, pergolas, or along fences to create beautiful garden features.

Pruning and Training: Key Tips

1. Research and Planning: Before pruning or training a plant, research its specific requirements and the best methods for achieving your goals.

2. Use the Right Tools: Pruning requires sharp, clean tools to make precise cuts and minimize damage to the plant.

3. Sanitation: After pruning diseased plant parts, clean your tools to prevent the spread of diseases to other plants.

4. Timing Matters: Prune at the appropriate time for each plant species. Some plants are best pruned in spring, while others benefit from fall or winter pruning.

5. Monitor Growth: Regularly check on the growth of your plants to ensure they're conforming to the desired shape or size.

Pruning and training are not only practices for shaping and managing your garden but also opportunities to connect with your plants, understand their growth habits, and engage in the creative aspect of gardening. When done thoughtfully and skillfully, these techniques contribute to a garden that is not only beautiful but also productive and healthy.

Maintaining a Healthy Garden Ecosystem

A garden is not just a collection of plants; it's an ecosystem teeming with life, from the tiniest microorganisms in the soil to the buzzing pollinators above. As a gardener, it's your role to steward this living system, ensuring its vitality, balance, and sustainability. To maintain a healthy garden ecosystem, you need to foster biodiversity, promote natural processes, and minimize harmful impacts on the environment.

1. Plant Diversity:

Diverse plantings are the cornerstone of a healthy garden ecosystem. Different species attract various pollinators, beneficial insects, and wildlife, creating a harmonious environment where each organism finds its role. By selecting a variety of plants, you can create niches for an array of life forms. Perennials, annuals, shrubs, trees, and groundcovers all contribute to a balanced ecosystem.

2. Native Plants:

Including native plants in your garden is especially important. These plants have evolved alongside local wildlife and are often the preferred food sources for native insects and birds. They can help restore and maintain the natural balance of your region's ecosystem.

3. Pollinator-Friendly Plants:

Supporting pollinators is crucial for a thriving garden ecosystem. Plant nectar-rich flowers to attract bees, butterflies, and other pollinators that play a vital role in plant reproduction.

4. Beneficial Insects:

Encourage the presence of beneficial insects like ladybugs, lacewings, and parasitic wasps. These predators help control pest populations naturally, reducing the need for chemical pesticides.

5. Organic Gardening:

Embrace organic gardening practices to protect the soil, water, and air from harmful chemicals. Organic gardening promotes the health of beneficial soil organisms and minimizes disruptions to the ecosystem.

6. Composting and Soil Health:

Maintaining healthy soil is at the heart of a thriving garden ecosystem. Composting enriches the soil with organic matter, fostering the growth of beneficial microorganisms. A healthy soil ecosystem supports strong plant growth and aids in disease and pest resistance.

7. Water Conservation:

Practicing responsible water management is essential for maintaining a garden ecosystem. Overwatering can wash away nutrients and disrupt soil life. Consider installing rain barrels and using drip irrigation or soaker hoses to water efficiently.

8. Avoiding Excessive Tilling:

Excessive tilling can disrupt the soil's structure and harm beneficial soil organisms. Minimize soil disturbance to maintain a stable, healthy ecosystem beneath the surface.

9. Wildlife-Friendly Features:

Create features like birdhouses, birdbaths, and butterfly feeders to attract and support wildlife in your garden. Wildlife plays an important role in pollination and pest control.

10. Pest and Weed Management:

Opt for integrated pest management (IPM) techniques to address garden challenges. IPM emphasizes prevention, monitoring, and the use of natural solutions, reducing the need for chemical treatments.

11. Companion Planting:

Practice companion planting to enhance the relationships between plants and create a balanced ecosystem. For example, planting marigolds near tomatoes can deter aphids, while basil can help protect tomatoes from hornworms.

12. Mulch:

Mulch garden beds to suppress weeds, conserve moisture, and moderate soil temperature. Organic mulch eventually breaks down, enriching the soil and supporting a diverse ecosystem within it.

13. Responsible Plant Selection:

Research the plants you introduce to your garden to ensure they are not invasive species that could disrupt the local ecosystem.

14. Pruning and Training:

Properly prune and train your plants to maintain healthy growth and structure while providing shelter and habitat for beneficial wildlife.

Maintaining a healthy garden ecosystem is a dynamic and rewarding practice. It connects you to the rhythms of nature and enriches your gardening experience. By creating a balanced and sustainable garden, you not only foster the well-being of your plants but also support the countless organisms that make your garden a living, thriving ecosystem.

Seasonal Maintenance and Care

A successful garden is a reflection of your care and attention throughout the year. Just as the seasons change, so do the needs of your garden.

Effective seasonal maintenance and care are crucial for keeping your garden healthy, vibrant, and productive. By understanding and addressing the unique demands of each season, you can ensure your garden thrives in every phase of the year.

Spring: Renewal and Growth

Spring is a time of renewal, with your garden awakening from its winter slumber.

Key tasks for spring maintenance include:

1. Cleaning and Prepping: Clear debris, fallen leaves, and dead plants from the previous year. Rake and aerate the soil to prepare it for new plantings.

2. Pruning: Trim back dead or damaged branches and shrubs to encourage fresh growth.

3. Soil Testing: Test your soil's pH and nutrient levels to determine if amendments are needed.

4. Planting: Sow seeds and transplant seedlings for annuals and vegetables. Plant perennials, shrubs, and trees during this active growing season.

5. Weed Control: Stay vigilant against early-season weeds, removing them before they take hold.

6. Fertilizing: Apply organic or balanced fertilizers to give your plants a nutrient boost for the growing season.

7. Pest Prevention: Implement strategies to deter common garden pests like aphids and slugs.

Summer: Vigilance and Hydration

Summer is a time of growth and warmth, but it also poses challenges like heat and pests. Seasonal maintenance in summer includes:

1. Watering: Provide consistent and deep watering, especially during hot spells, to keep your plants hydrated.

2. Mulching: Maintain mulch to conserve moisture, regulate soil temperature, and suppress weeds.

3. Deadheading: Remove spent flowers to encourage continuous blooming in annuals and perennials.

4. Pest Control: Continue monitoring for pests and apply natural remedies or interventions as needed.

5. Harvesting: Gather your garden's bounty regularly to encourage further production.

6. Pruning: Light pruning can help control the size and shape of plants while improving airflow.

7. Shade and Sun Protection: Provide shade for delicate plants and protect them from harsh midday sun.

Fall: Transition and Preparation

Fall signifies the transition from growth to dormancy, and maintenance tasks shift accordingly:

1. Harvest: Gather the last of your vegetables and fruits before frost arrives.

2. Cut Back: Prune dead or overgrown branches and stems, and remove annual plants that have completed their lifecycle.

3. Planting: Consider planting fall crops and bulbs for the next year. Fall is also an excellent time for planting trees and shrubs.

4. Clean Up: Remove fallen leaves, spent plants, and debris from the garden.

5. Composting: Add organic matter to the soil to enhance its fertility over the winter.

6. Mulching: Apply fresh mulch to protect plants from winter cold and suppress weeds.

7. Protect Tender Plants: Cover or move tender plants indoors to safeguard them from frost.

Winter: Rest and Reflection

Winter is a season of rest for your garden, but it's also a time for reflection and planning:

1. Plan Ahead: Review your garden's performance and consider changes for the upcoming year. Order seeds and plan your garden layout.

2. Tool Maintenance: Clean, sharpen, and maintain your gardening tools, ensuring they are ready for the next growing season.

3. Indoor Gardening: If you enjoy indoor plants, continue caring for them and consider adding new ones to your collection.

4. Composting: Continue composting kitchen scraps and garden waste to prepare for spring.

5. Bird Feeding: Provide food, water, and shelter for wintering birds to encourage biodiversity in your garden.

CHAPTER FOUR

Nurturing the Fruits of Your Labor

As the days grow longer and the sun's warmth embraces your garden, a culmination of your hard work and dedication draws near.

The time for harvesting has arrived, marking a pivotal and rewarding moment in your gardening journey. It's a time when the lush, ripe offerings of your garden burst forth, filling your baskets with the vibrant, earthy treasures you've nurtured from tiny seeds and tender seedlings.

Yet, harvesting is more than just the act of gathering crops; it's a celebration of nature's abundance and a reflection of your deep connection to the soil, seasons, and the timeless rhythms of life.

Here, we delve into the art and science of reaping the rewards of your garden, from plucking the ripest tomatoes and crisp lettuces to picking the juiciest peaches and crunchiest apples. But our exploration goes far beyond the act of harvesting itself. It extends into the realms of preservation, storage, and culinary creativity, where the flavors and aromas of your garden are transformed into delectable dishes, pickles, jams, and storable delights that capture the essence of each season.

This book is a guide, a companion, and a source of inspiration as you navigate the rich, fulfilling chapters of your gardening adventure. Whether you're a seasoned gardener with years of experience or a novice just beginning to cultivate your green thumb, there is something here for you. This is a comprehensive and insightful resource that takes you through the entire process of reaping, preserving, and savoring the fruits of your labor.

In the chapters to come, we'll journey together through the bountiful world of the harvest, exploring the nuances of timing, techniques, and tools that ensure you gather the finest produce. We'll discover the secrets of storing your harvest to maintain its freshness and flavor well beyond the growing season, whether you're a fan of canning, freezing, or root cellaring. And we'll venture into the heart of culinary creativity, where the garden's offerings are transformed into mouthwatering recipes that celebrate the distinct tastes of each season.

But "Growing your own food" is not just a practical manual; it's a testament to the joys of tending to the soil and the satisfaction of nurturing a garden from seed to table.

It's a tribute to the reverence and respect for the natural world that gardeners cultivate, as well as the profound sense of connection to the seasons that unfolds with each new planting and harvest.

It's a reminder that, in the ever-accelerating pace of modern life, there is solace and wisdom to be found in the simplicity of tilling the earth, waiting for the first shoots to emerge, and ultimately reaping the harvest of your labor.

So, as you embark on this journey through the pages of "Growing your own food," prepare to embrace the magic of the garden's annual bounty.

Let the wisdom within these chapters be your guide as you nurture the fruits of your labor, transforming them into delicious dishes, preserving the essence of each season, and savoring the rich flavors and aromas that only a garden can provide. Welcome to a world where the earth's gifts are yours to cultivate, cherish, and share with those you hold dear.

Recognizing Harvest-Ready Crops

The act of harvesting is one of the most rewarding moments in a gardener's journey. It's a time when you get to enjoy the literal fruits of your labor, plucking vegetables, fruits, and herbs at the peak of their flavor and nutritional value.

However, the question that often arises is, "When is the right time to harvest?" Recognizing the moment when your crops are at their prime for harvesting is both a science and an art, and it can greatly affect the quality of your yield. Let's explore how to recognize when various crops are ready to be harvested.

1. Visual Clues:

Visual cues are among the most reliable indicators for many crops.

Here's how to recognize when some of your favorite vegetables and fruits are ready:

- **Tomatoes:** Ripe tomatoes often have a vibrant color, whether it's deep red, orange, or another hue depending on the variety. They should also feel firm but slightly give when gently squeezed.

- **Cucumbers:** Pick cucumbers when they reach a deep green color and are firm to the touch. Avoid overripe cucumbers that turn yellow and become mushy.

- **Bell Peppers:** Wait until bell peppers have reached their full color potential, whether it's green, red, yellow, or another shade. They should be firm and crisp.

- **Zucchini and Squash:** Harvest zucchinis and summer squash when they're around 6-8 inches long and their skin is still tender. Overly large squash can be tough and less flavorful.

- **Beans:** Harvest green beans when they're firm and snap easily. They should also be free of bulges or blemishes.

- **Eggplants:** Ripe eggplants have a glossy sheen, firm skin, and vibrant color. Press your thumb into the skin; if it springs back, it's ready.

2. Taste Test:

For many crops, the best way to tell if they're ready is to taste them. This is particularly true for fruits and herbs. If you're growing berries, such as strawberries or raspberries, try one to see if it's sweet and flavorful.

When it comes to herbs like basil, thyme, or cilantro, the aroma and taste will guide you.

3. Fruit and Stem Separation:

For fruits like melons, cantaloupes, and watermelons, they should easily separate from the vine with a gentle tug. If they're clinging to the stem, they may not be ripe yet. Conversely, for crops like corn, you can check their readiness by piercing a kernel; if a milky substance emerges, they're ready to harvest.

4. Leafy Greens:

Leafy greens, such as lettuce, kale, and spinach, can be harvested when the leaves are young and tender. You can start by picking the outer leaves and allowing the inner ones to continue growing. If the greens begin to bolt, forming flowers or seeds, they may become bitter and less palatable.

5. Root Vegetables:

For crops like carrots and beets, it's a bit more challenging to assess readiness. Gently pull one from the soil to check its size. If it's reached a desirable size for your culinary needs, it's ready to harvest.

6. Onions and Garlic:

Onions and garlic are ready for harvest when their tops start to turn yellow and fall over. At this stage, the bulbs are mature and can be lifted from the soil for curing.

7. Herbs:

Herbs like basil, cilantro, and parsley can be harvested when they've grown enough to provide a substantial amount for your culinary needs. Regular harvesting can also encourage bushier, more productive growth.

8. Fruit Trees:

Fruit trees require close attention to their specific harvest times. Varieties like apples and pears should be picked when they have reached their full color and easily come off the tree with a gentle twist. Peaches and nectarines should also give slightly when you gently press them.

Remember that the timing of harvest can vary depending on your local climate, the specific variety of the crop, and growing conditions. Regularly inspect your plants and taste-test as needed to ensure you pick your crops at the perfect moment, maximizing flavor and nutritional value.

By mastering the art of recognizing when your crops are at their prime, you can savor the true essence of your garden's bounty.

Harvesting Techniques for Different Vegetables and Fruits

The art of harvesting goes beyond just picking vegetables and fruits at the right time. Different crops require distinct techniques to ensure they are harvested properly, preserving their flavor and quality. From gentle twists to delicate snips, let's explore the techniques for harvesting a variety of vegetables and fruits to maximize the delicious rewards of your garden.

1. Tomatoes:

Tomatoes should be harvested when they reach their peak ripeness, which is typically indicated by their vibrant color and a slightly soft feel. To pick them, gently twist the tomato where the stem connects to the plant. This technique helps avoid damage to the fruit. Be cautious not to pull on the tomato too hard, as this can damage the plant and other fruits.

2. Cucumbers:

Cucumbers should be harvested when they are firm, have a deep green color, and are about 6-8 inches in length. Use scissors or a sharp knife to cut the cucumber from the vine rather than pulling or twisting, which can damage the plant. It's essential to be gentle, as excessive force can harm the vine and nearby cucumbers.

3. Bell Peppers:

Bell peppers can be harvested when they have reached their full color potential, whether it's green, red, yellow, or another shade, and when they are firm to the touch.

Use scissors or garden shears to cut the stem just above the pepper to avoid damaging the plant.

4. Zucchini and Summer Squash:

Zucchinis and summer squash are ready for harvest when they are about 6-8 inches long and still have tender skin. Use scissors or a sharp knife to cut the squash from the vine without pulling. Be sure not to leave a stub on the plant, as this can attract pests and diseases.

5. Beans:

Green beans should be picked when they are firm, crisp, and about 4-6 inches in length. Use scissors or garden shears to cut the beans from the plant. Harvest them regularly to encourage more production.

6. Eggplants:

Eggplants are ripe when they have a glossy sheen, firm skin, and a vibrant color. To harvest, use scissors or a knife to cut the eggplant from the plant. Be gentle to avoid damaging the plant or other nearby eggplants.

7. Berries:

Berries like strawberries, raspberries, and blueberries should be gently plucked from the plant when they are fully ripe and easily detach from the stem with a slight twist or tug. Be careful not to crush or bruise the delicate fruits during harvest.

8. Lettuce and Leafy Greens:

Leafy greens are best picked when they are young and tender. Use scissors or garden shears to cut the outer leaves, allowing the inner ones to continue growing. Harvest regularly to keep the plants productive.

9. Root Vegetables:

For crops like carrots and beets, use a garden fork or a hand trowel to carefully loosen the soil around the root before pulling it out. This technique minimizes damage to the roots and allows for easier harvesting.

10. Onions and Garlic:

Onions and garlic should be pulled from the soil when their tops have turned yellow and fallen over. Lift them gently to avoid damage during harvest.

11. Fruit Trees:

Fruit trees such as apples, pears, peaches, and plums require gentle twisting and lifting. For apples and pears, turn the fruit slightly and lift upward until it easily detaches from the tree. For stone fruits like peaches and plums, give them a slight twist, and they should come free with little effort.

12. Herbs:

Herbs like basil, cilantro, and parsley can be harvested by

snipping or pinching off individual leaves or stems with scissors or your fingers. Regular harvesting can encourage more abundant growth.

Understanding the specific harvesting techniques for different crops ensures that your hard-earned garden bounty is gathered and enjoyed at its finest.

Practicing gentleness and care while using the right tools is key to preserving the quality of your vegetables and fruits, ultimately enhancing your culinary experience and the satisfaction of savoring the fruits of your labor.

Food Preservation Methods: Canning, Freezing, and Drying

Preserving the abundance of your garden or the seasonal offerings of local markets is a time-honored tradition that allows you to savor the flavors of your favorite fruits and vegetables throughout the year.

Three primary methods of food preservation—canning, freezing, and drying—offer different ways to extend the shelf life of your harvest while maintaining the taste, nutrition, and quality of your produce.

1. Canning:

Canning is a popular food preservation method that involves heat-processing foods in sealed containers to create an airtight seal that prevents spoilage. Two main canning methods exist: water bath canning and pressure canning.

- **Water Bath Canning:** This method is suitable for preserving high-acid foods like fruits, pickles, jams, and tomatoes. The process involves immersing jars of prepared food in boiling water for a specified period. As the jars cool, a vacuum seal is formed, preventing the growth of spoilage organisms.

- **Pressure Canning:** Pressure canning is necessary for low-acid foods like vegetables, meats, and poultry. It involves using a pressure canner to achieve the high temperatures needed to kill harmful bacteria, such as botulism spores. The sealed jars preserve these low-acid foods safely.

2. Freezing:

Freezing is a straightforward and effective method for preserving many types of produce. This method retains the natural color, flavor, and nutritional value of your food.

Here's how to freeze your harvest:

- **Blanching:** Many vegetables benefit from blanching before freezing. This involves briefly immersing them in boiling water, followed by a quick chill in ice water. Blanching helps retain the color, flavor, and texture of the vegetables while killing enzymes that can lead to spoilage.

- **Packaging:** After blanching, package your vegetables in airtight containers or freezer bags, removing as much air as possible to prevent freezer burn. Label and date the packages for easy identification.

- **Fruits:** Most fruits can be frozen without blanching. Simply wash, peel, and slice as needed, then package in airtight containers or bags.

3. Drying:

Drying, or dehydrating, removes the moisture from food, making it less susceptible to spoilage and degradation. This method is especially suitable for fruits, herbs, and some vegetables. There are various ways to dry food:

- **Sun Drying:** This traditional method involves spreading food out in the sun to allow the sun's heat and air circulation to dry the produce. It works well for fruits, herbs, and tomatoes.

- Oven Drying: In an oven, set to low temperatures, food can be dried by placing it on racks or trays. This method works for herbs, fruits, and some vegetables.

- Dehydrator: A food dehydrator offers precise temperature control and good air circulation. It's an efficient and consistent way to dry a wide range of foods, from fruits and vegetables to jerky and herbs.

Proper storage is essential for all preserved foods. Store canned goods in a cool, dark place, away from direct sunlight. Frozen produce should remain in the freezer, while dried foods should be kept in airtight containers and stored in a cool, dark location to maintain freshness.

Food preservation allows you to enjoy the flavors of summer in the depths of winter, and it's an excellent way to reduce waste and make the most of your garden's bounty. Whether you're canning, freezing, or drying, each method offers unique benefits, allowing you to savor the flavors and nutrition of your homegrown produce all year long.

Sharing Your Harvest and Community Gardening

The joy of gardening isn't limited to the satisfaction of nurturing your own plants and savoring the bountiful harvest; it's also about the community and connection that grows around the garden. Sharing the fruits of your labor and participating in community gardening initiatives can be equally rewarding, creating a sense of togetherness, purpose, and a deeper connection to the land. Here, we'll explore the value of sharing your harvest and the benefits of community gardening.

Sharing Your Harvest

1. Generosity and Connection: One of the most beautiful aspects of gardening is the opportunity to share the abundance with family, friends, and neighbors. By giving away your surplus vegetables and fruits, you not only strengthen your bonds with others but also create a sense of generosity and goodwill in your community.

2. Reducing Food Waste: Sharing your harvest helps minimize food waste. Produce that might otherwise go unused or spoil in your kitchen can find a home where it's appreciated and enjoyed.

3. Educational Opportunity: Sharing your harvest offers an excellent chance to educate others about the importance of homegrown, fresh produce. You can share gardening tips, cooking recipes, and the knowledge of growing and nurturing plants.

4. Healthy Eating: By gifting your harvest, you encourage others to incorporate more fresh, locally grown produce into their diets, promoting healthier eating habits and supporting local agriculture.

5. Community-Building: Acts of sharing create a sense of community and solidarity. Neighbors who receive your harvest may reciprocate with their own surplus, fostering a culture of mutual support and exchange.

Community Gardening

1. Shared Resources: Community gardens provide access to shared resources, like tools, water, and space, making gardening more accessible to those without their own land.

2. Diverse Expertise: Community gardens often attract people with various gardening backgrounds and knowledge.

This diversity can provide valuable insights and foster learning and skill-sharing.

3. Collective Efforts: Gardening in a group can make large projects more manageable. Community gardeners can collaborate on tasks like building compost bins, erecting trellises, or constructing raised beds.

4. Enhanced Environmental Stewardship: Community gardens encourage environmentally friendly practices like composting, organic gardening, and water conservation. They can contribute to local sustainability and wildlife habitat creation.

5. Social Connections: Community gardening is not just about growing plants; it's also about growing friendships. Working side by side with fellow gardeners offers social interaction and a sense of belonging.

6. Educational Opportunities: Many community gardens host workshops, classes, and events that promote gardening skills, healthy eating, and environmental awareness.

7. Food Security: Community gardens can play a role in addressing food security issues, providing fresh produce to those who may have limited access to it.

8. Cultural Exchange: Community gardens often reflect the diverse backgrounds of their participants. This offers an opportunity for cultural exchange, learning about different cuisines and gardening traditions.

9. Urban Green Spaces: Community gardens can transform vacant urban lots into green oases, improving the quality of life in densely populated areas.

By sharing your harvest and participating in community gardening, you not only enjoy the tangible rewards of homegrown produce but also experience the intangible benefits of community and connection. It's a way to nurture not just your garden but also the bonds between people, fostering a sense of unity and purpose as you collectively tend to the Earth and its bounty.

Sustainability and Reducing Food Waste

Sustainability and reducing food waste are critical components of responsible and conscious gardening and food production. In a world where food security, environmental conservation, and resource management are paramount, understanding how to cultivate and consume food while minimizing waste is essential. Let's explore the importance of sustainability in gardening and practical strategies for reducing food waste.

The Importance of Sustainability

1. Resource Conservation: Sustainable gardening focuses on the efficient use of resources such as water, energy, and soil. It minimizes the environmental footprint of gardening practices by making the most of available resources.

2. Biodiversity: Sustainable gardening practices prioritize the preservation and enhancement of biodiversity. This approach ensures that a variety of plants and beneficial insects thrive in your garden, contributing to a healthy and resilient ecosystem.

3. Soil Health: Sustainable gardening seeks to improve and protect soil health. This involves practices like composting, mulching, and minimal soil disruption, which support healthy microbial communities and robust plant growth.

4. Water Efficiency: Sustainable gardening minimizes water waste by using techniques like drip irrigation and rainwater harvesting, as well as selecting drought-tolerant plants.

5. Chemical Reduction: Reducing the use of chemical pesticides and synthetic fertilizers is a fundamental aspect of sustainability. Organic and natural methods of pest control and soil enrichment are favored.

Reducing Food Waste:

1. Harvest Timing: Harvest your produce at its peak ripeness, ensuring that it reaches your table in the best possible condition. Overripe or overly mature fruits and vegetables are more likely to go to waste.

2. Preservation: Use food preservation techniques such as canning, freezing, and drying to extend the shelf life of your harvest. These methods enable you to enjoy your garden's bounty long after the growing season ends.

3. Composting: Compost plant waste, including kitchen scraps and garden trimmings, to recycle organic matter and enrich your soil. Composting reduces landfill waste and supports a sustainable nutrient cycle.

4. Mindful Storage: Store your produce properly to extend its freshness. Some items should be kept in a cool, dark place, while others do better in the refrigerator or freezer. Proper storage helps reduce spoilage.

5. Regular Inspection: Regularly check your stored produce for signs of spoilage or damage. Remove any spoiled items to prevent them from affecting others.

6. Preserve Surplus: When you have a surplus of a particular crop, consider sharing it with friends, family, or neighbors, or donate it to local food banks or community organizations.

7. Meal Planning: Plan your meals to use up fresh ingredients before they spoil. This reduces the likelihood of food waste in your kitchen.

8. Versatile Cooking: Be creative in the kitchen by using all parts of a plant, including stems, leaves, and peels, in your recipes. You can make use of scraps for broths or sauces.

9. Repurposing: Transform leftovers into new dishes to reduce food waste. For example, stale bread can become croutons or breadcrumbs, and vegetable scraps can be used for homemade vegetable broth.

10. Educate and Advocate: Share your knowledge and passion for reducing food waste with friends, family, and your community. Advocate for policies that promote food recovery and waste reduction.

Sustainability and reducing food waste are essential principles for responsible gardening and conscious consumption. By adopting these practices, you not only contribute to a healthier and more sustainable environment but also enjoy the many benefits of fresher, tastier, and more environmentally friendly food. Your garden can be a symbol of both sustainability and responsible stewardship of the Earth's resources.

Lessons Learned and Planning for Next Season

Every gardening season is a journey filled with growth, discoveries, challenges, and triumphs. As the final harvest is gathered, and the first frost approaches, it's the perfect time to reflect on the lessons learned and start planning for the next season. These moments of reflection and planning are the cornerstones of continuous improvement and success in your garden.

Lessons Learned

1. Plant Selection: Reflect on the success and challenges of the plant varieties you chose. Were there specific cultivars that thrived in your climate and soil? Were there any that struggled or were susceptible to pests and diseases?

2. Timing: Consider the timing of your plantings. Did you start seeds or transplant seedlings at the right time? Did you stagger plantings for a continuous harvest, or did you experience gluts and gaps in production?

3. Soil Health: Assess the condition of your garden soil. Did you amend it properly before planting? Were there any issues with nutrient deficiencies or pH levels? A soil test can provide valuable insights.

4. Pest and Disease Management: Reflect on your strategies for pest and disease management. Were there any significant issues, and how did you address them? Did you implement companion planting or other natural pest control methods effectively?

5. Irrigation: Consider your watering practices. Did you provide consistent and adequate moisture for your plants?

Did you use water-efficient techniques, such as drip irrigation or soaker hoses?

6. Harvest Timing: Evaluate your harvest timing. Did you consistently pick fruits and vegetables at their peak ripeness, or did you encounter underripe or overripe produce?

7. Crop Rotation: Reflect on your crop rotation strategy, if you implemented one. Crop rotation can help prevent soil-borne diseases and improve soil health.

8. Composting and Organic Matter: Assess your composting practices. Did you consistently add organic matter to your garden to enrich the soil, improve structure, and support beneficial microbial life?

Planning for Next Season

1. Crop Selection: Use the knowledge gained from this season to make informed choices about the crops you want to grow next year. Consider experimenting with new varieties and heirlooms to diversify your garden.

2. Garden Layout: Plan the layout of your garden beds. Ensure you practice proper crop rotation to prevent the build-up of soil-borne diseases and maintain healthy soil.

3. Soil Preparation: Prepare your garden soil by adding compost and other organic matter. Conduct a soil test to determine any nutrient deficiencies and adjust the pH if necessary.

4. Seed Starting: If you start your plants from seeds, create a seed-starting schedule to ensure timely germination and healthy seedlings. Adequate indoor lighting and temperature control are key.

5. Pest and Disease Management: Develop a holistic pest and disease management plan. Implement preventive measures, such as companion planting and beneficial insect attraction, and have solutions ready for specific problems.

6. Water Management: Design a water-efficient irrigation system, whether it's drip lines, soaker hoses, or rain barrels. Proper watering helps conserve resources and supports healthy plants.

7. Composting: Continue composting kitchen scraps and garden waste to enrich your soil and maintain a thriving garden ecosystem.

8. Succession Planting: Plan for succession planting to ensure a continuous supply of fresh produce throughout the growing season.

9. Education: Stay informed and educate yourself about gardening best practices, new techniques, and the latest developments in sustainable and organic gardening.

10. Community: If you're part of a gardening community, engage with fellow gardeners to share insights, support one another, and learn from their experiences.

By reflecting on your gardening season and planning for the next one, you're not only enhancing your chances of a successful harvest but also embracing the beautiful cycle of growth and renewal that gardening offers. Each year provides new opportunities to learn, experiment, and connect with the natural world, enriching your life in countless ways.

Resources for Ongoing Learning and Inspiration

Gardening is a lifelong journey of discovery and cultivation. To keep your gardening skills sharp and your enthusiasm blooming, it's essential to have access to valuable resources for ongoing learning and inspiration. Whether you're a seasoned gardener looking to refine your techniques or a beginner eager to dive into the world of horticulture, these resources can be your companions on this green and growing path.

1. Gardening Books:

Gardening books are timeless treasures. They offer in-depth knowledge, practical advice, and a wealth of inspiration. Whether you're interested in specific plant families, gardening styles, or design principles, there's a gardening book to guide and inspire you.

2. Online Gardening Communities:

Joining online gardening communities can connect you with fellow enthusiasts, experts, and experienced gardeners. Websites, forums, and social media groups provide a platform for sharing insights, asking questions, and seeking advice from a global network of gardeners. Popular platforms like GardenWeb, Houzz, and Reddit's gardening subreddit are great places to start.

3. Gardening Magazines:

Gardening magazines are a source of ongoing inspiration. They feature beautiful photographs, informative articles, and the latest trends in gardening. Publications like "Fine Gardening," "The English Garden," and "Garden Design" offer a wealth of ideas and tips.

4. Local Gardening Clubs and Associations:

Many regions have local gardening clubs and associations where members share knowledge, host events, and even garden together. These community-based organizations provide hands-on learning opportunities, from plant swaps to garden tours. Check out local horticultural societies and master gardener programs in your area.

5. Botanical Gardens and Arboretums:

Botanical gardens and arboretums are living showcases of plant diversity and design. Visiting these institutions not only offers inspiration but also educational programs, workshops, and gardening exhibitions. Many botanical gardens have online resources and events as well.

6. Garden Tours and Open Houses:

Participating in garden tours and open houses provides the opportunity to see different gardening styles, plant combinations, and landscape design principles in action. You can gain ideas and insights from visiting well-established gardens in your area.

7. Online Courses and Webinars:

Numerous universities, extension services, and gardening organizations offer online courses and webinars on various horticultural topics. These educational resources allow you to delve into specific areas of interest and gain knowledge from the comfort of your own home.

8. YouTube Channels and Podcasts:

YouTube gardening channels and podcasts are engaging platforms for learning and inspiration. Experts and passionate gardeners share their knowledge, tips, and experiences through videos and audio content. Some popular channels and podcasts include "Gardeners' World," "Epic Gardening," and "A Way to Garden."

9. Gardening Apps:

There are a plethora of gardening apps available for both

beginners and experienced gardeners. These apps can help you identify plants, manage garden tasks, and access gardening advice on your mobile device.

10. Public Libraries:

Don't overlook your local public library as a valuable resource for gardening books, magazines, and educational materials. Libraries often host gardening events and workshops as well.

Embrace the wealth of resources available to you as a gardener. From the wisdom shared in books to the connections formed in local clubs and the ever-expanding world of online information, these resources will keep your gardening journey vibrant, engaging, and continuously evolving.

So, go ahead and dig into the world of gardening knowledge and inspiration, and watch your green passions flourish.

CONCLUSION

In the world of gardening and growing your own food, each season marks the turning of a new leaf, both figuratively and literally. As we reach the conclusion of this journey, it's evident that growing your own food is not merely a task but a transformational experience that touches every facet of life. It's a labor of love, a communion with nature, a source of sustenance, and a classroom for life's most valuable lessons.

The benefits of growing your own food are abundant, from the exquisite flavors of homegrown produce to the profound sense of self-reliance and resilience. In the process, we learn patience as we nurture seeds into bountiful harvests, understanding as we delve into the intricacies of soil health and plant biology, and gratitude as we appreciate the precious gift of nature's abundance.

But the journey of growing your own food is not one traveled alone. It's a shared path that connects you with your community, whether through swapping seeds with neighbors, sharing your garden's bounty, or participating in local gardening clubs. The support and inspiration derived from fellow gardeners create a sense of unity and a shared commitment to sustainability, reducing food waste, and cherishing the earth's gifts.

As you till the soil, plant seeds, and watch your garden flourish, you become part of a timeless cycle, in tune with the seasons and the natural world. You also contribute to environmental stewardship, practicing sustainable agriculture, and reducing your carbon footprint.

In the end, growing your own food is not just about the final harvest; it's about the journey itself, filled with learning, curiosity, and the simple pleasure of being connected to the earth. So, as we conclude this exploration of growing your own food, remember that the journey never truly ends. It's an ongoing odyssey, a celebration of life, and a testament to the human connection to the land.

As each seed takes root and each plant unfurls its leaves, it's a reminder that with care, patience, and a deep respect for the earth, we can cultivate our own sustenance and continue to reap the many rewards of a life deeply intertwined with the world of gardening.

www.ingramcontent.com/pod-product-compliance
Lightning Source LLC
Chambersburg PA
CBHW062314290526
45794CB00005B/1802